To the User

This book is not a course, but it will be used in a contex........... learning a diversified curriculum of skills, knowledge and positive attitudes about science. Its main purpose is to provide a first-aid or survival kit of lively ideas for teachers, who may not be scientists, to use in emergencies.

Who are these teachers?

Primary school teachers just looking for new ideas.
Secondary teachers seeking projects for less-able students.
Supply teachers in need of 'instant' science lessons.
Head teachers who cover for absent colleagues.
Student-teachers desperate for easy-to-prepare material.
Parents who are fortunate enough to discover this book.
Children who may be trusted to educate themselves.

If these persons want practical suggestions on the fringe of but not incompatible with the contents of textbooks, they need look no further . . .

Many of the projects can succeed with the raw materials of routine classroom work, although there is sometimes a need for minimal extras, such as a sweet or pickles jar, a stop-watch (even, in one case, a few grapes and a bottle of cheap lemonade).

Try to find time to test the ideas before expecting children to do them — then you will be able to anticipate difficulties that children may have.

I believe it is important for science to be enjoyable, therefore I have injected a generous dose of fun into this book.

© Collins Educational

First published in 1988 by Holmes McDougall Ltd., Edinburgh

This edition published in 1990 by Collins Educational, London and Glasgow

Reprinted 1992

ISBN 0 00 329474 9

All rights reserved. No part of this publication may be reproduced, stored in a retrieval system, or transmitted in any form or by any means, electronic, mechanical, recording or otherwise, without the prior permission of the publishers, except those assignment and resource pages which are clearly marked.

Printed by Martin's The Printers Ltd., Berwick on Tweed

The Guiding Principles

Science is working and thinking in certain ways to gain knowledge that is useful for understanding how nature works and for practical and economical uses in problem-solving and technology.

Science involves the following:
making observations
asking questions
making comparisons
sorting out ideas
making measurements
recording in words
recording in numbers
making predictions (based on knowledge)
doing fair tests

Doing a fair test means taking into account those various factors (variable) that may affect the results of a test . . . Questions to be answered by observation and fair testing should be kept clear and simple.

Nature includes the whole of the physical world, non-living as well as living things. Technology is about using reliable scientific knowledge to make life on earth more interesting, comfortable and secure — with the least financial expense and least waste of resources. (The ideal of a decent quality of existence should be assumed.) A problem exists when there is no apparent answer to a question.

My ideas are inspired and guided by these principles.

HOMEMADE HOLLYWOOD

What you need
A4 size paper
Nail about 10cm long

Tear off a strip of A4 size paper, about 6cm wide. Fold the strip in half and open out the paper again.

Copy portrait A in the position shown in the picture. Fold the paper and trace portrait A – while making slight changes to the details – to obtain B, which will be on top of A.

Roll B over a nail. This is to make the paper springy.

Remove the nail. Then roll this springy part of the paper around an end of a pencil.

Rest the paper on a firm surface. Grip its folded edge. Use your free hand to slide the pencil up and down.

Sliding the pencil causes A and B to be exposed alternately.

<u>Practise</u> (if necessary making fresh drawings) until you get an amazing impression of a moving picture.

How does this scientific toy help you to understand how TV and cinema pictures are made?
Who amongst your friends can make the most entertaining movie?

HOLD YOUR OWN FILM FESTIVAL !!!

PLAN THOUGHTFULLY BECAUSE ONLY CERTAIN KINDS OF IDEAS WILL WORK EFFECTIVELY.

© Collins Educational Science Teachers

You may photocopy this page for use within the classroom.

MINUTE MARATHON

What you need
A watch that times seconds.
CHILDREN CAN WORK IN GROUPS.

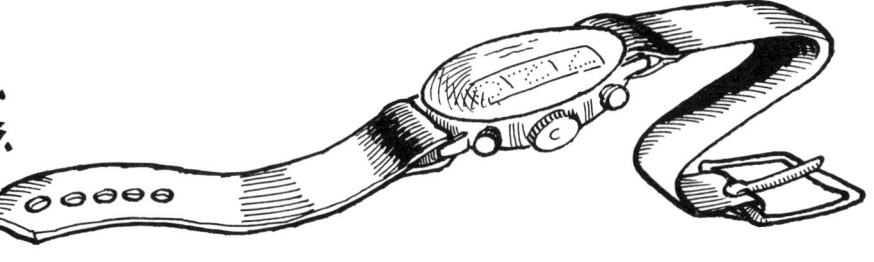

CAN YOU JUDGE A MINUTE?
Use a watch that counts seconds. At EXACTLY the beginning of a minute (WHEN THE SECOND HAND IS ON 12 OR THE INSTANT THAT THE MINUTE DIGIT CHANGES ON A DIGITAL WATCH) **close your eyes.**

Open your eyes after you have judged one minute.
WAS YOUR JUDGEMENT **FAST** OR **SLOW**?
After practising, do you get better at judging one minute?

MINUTE MARATHONS:–
In just one minute:
- HOW MANY TIMES CAN YOU SIGN YOUR NAME?
- CAN YOU PRINT OUT THE ALPHABET IN FULL?
- CAN YOU STOP BREATHING (HOLD YOUR BREATH)?
- HOW FAR CAN YOU RUN?
- HOW MANY PAPER CLIPS CAN YOU PICK UP, ONE BY ONE, FROM THE FLOOR?

INVENT OTHER MINUTE MARATHONS.

Cupid's Mysterious Letter.

What you need
A GOOD DICTIONARY.

Figure out what Cupid has to say...

Write out Cupid's message.

Create your own mysterious message in this style.

This is an example of ANAMORPHOSIS.
(Find out what it means.)

Think of uses for anamorphic writing, patterns and pictures.
(They are used for some road markings....)

© Collins Educational Science Teachers

You may photocopy this page for use within the classroom.

Paper Banger Technology

What you need
Tabloid newspapers.
Various scrap papers.

I generally use two thicknesses of a small (tabloid) size newspaper, to make a paper "BANGER".

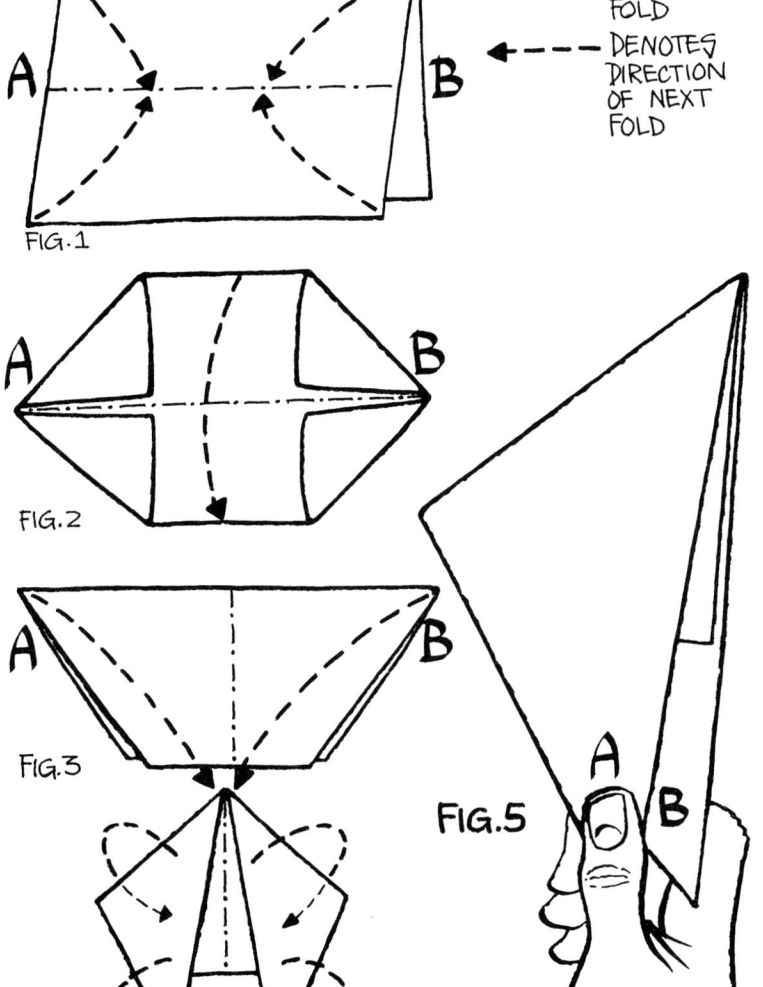

-·-·- DENOTES FOLD

◄---- DENOTES DIRECTION OF NEXT FOLD

Follow the diagrams up to figure 4. Fold the paper BACK, in half in to get the object shown in figure 5.
FIGURE 5 IS DRAWN ON A BIGGER SCALE — to make details clearer.

Hold the object tightly between points A and B — then swish it rapidly down through the air (rather like flicking a duster). The flap inside shoots out suddenly — making a loud BANG!

WHAT IS "THE BEST PAPER BANGER"?
1. The one that makes the loudest bang?
2. The one that works most times, without splitting?
3. The one that looks prettiest?
4. The one that is cheapest to make?
5. The one with all these qualities?

HOW DOES IT WORK?
WHY DO PEOPLE HEAR THE BANG?

Where on a piece of paper do you have to print the word 'BANG!' for this word to appear when the paper is folded and made to work as a banger?

© Collins Educational Science Teachers

You may photocopy this page for use within the classroom.

What you need

Mirror
Calculator (optional)

 Here are the capital letters of the alphabet:

A B C D E F G H I J K L M N O
P Q R S T U V W X Y Z

The letter 'A' has a vertical line of symmetry.

The letter 'B' has a horizontal line of symmetry.

If a mirror is held across the middle of a letter, with its bottom edge resting along a line of symmetry, the image in the mirror makes it look as if the letter is whole again.

Some letters have both vertical and horizontal forms of symmetry. Others, such as 'G' and 'Z' have neither of these forms.

© Collins Educational Science Teachers

WRITE CAPITAL LETTERS INSIDE THE APPROPRIATE CIRCLES

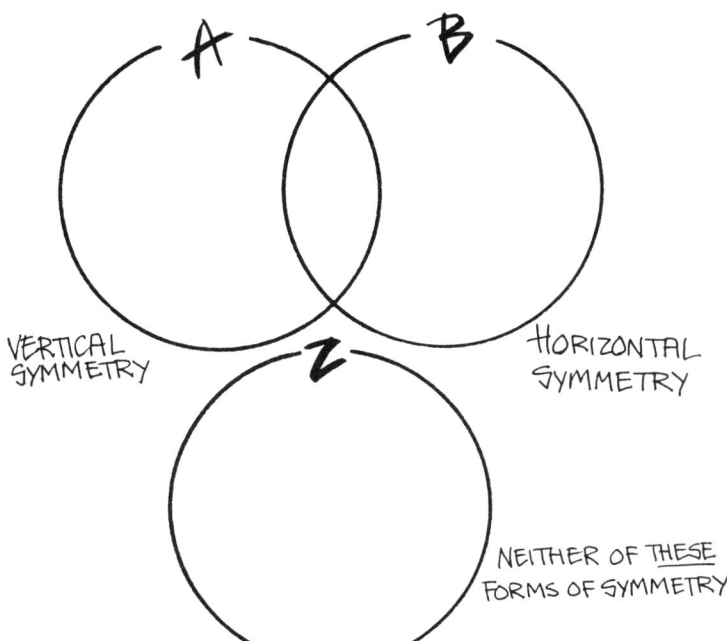

Which letters belong in the space where the circles overlap?

Print 'CHECKED OK' and 'ALL WRONG' on scraps of paper. Look at their reflections.
Try 'CARBON DIOXIDE' and 'CHOICE QUALITY'. Also try 'bob kicked pop' ....

How must you print 'TIMOTHY' for its mirror image to read normally?

USE THE IDEAS YOU HAVE DISCOVERED TO INVENT A TRICK.

Make words appear on a pocket calculator by turning certain numbers upside down.
Try printing the numbers 3045
37818
71077345

How many different 'calculated words' can you discover?

You may photocopy this page for use within the classroom.

twitchcraft ≡ NOT WITCHCRAFT

What you need

Selection of metal nuts
Cotton

A famous conjuror has a jewel on a thread made of gold silk. When he holds the dangling jewel – like a pendulum – over a girl's hand, the jewel starts to swing in a circle. If over a boy's hand, it starts to swing in a straight line. But what is truly amazing is that the jewel swings in these ways if a person who is not a conjuror holds the silk..... Some people think it is witchcraft, but this is not true.

If you know how the pendulum should swing (in a circle for a girl, in a straight line for a boy) you might discover that you can do this trick yourself – without making the pendulum swing on purpose!

The secret of success is in your mind. If your mind is impressed by a wish for the experiment to succeed, it makes your muscles move in ways so slight that you don't notice – but enough to make the pendulum swing in the different ways.

Make a pendulum by tying a metal nut on a cotton thread. Use it to try some tests amongst your friends – or you could try tests over boys' and girls' names written on pieces of paper. Experiment with different weights and lengths of the pendulum. WILL TESTS WORK IF THE PERSON HOLDING THE COTTON IS BLINDFOLDED?

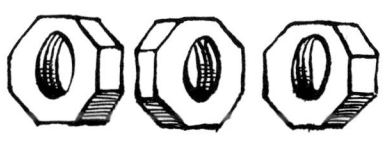

WHATEVER HAPPENS, REMEMBER THAT THERE IS NOTHING "WITCHY" ABOUT YOUR EXPERIMENTS. WHATEVER HAPPENS IS A SCIENTIFIC FACT.

© Collins Educational Science Teachers   You may photocopy this page for use within the classroom.

.... can minds be "read"?

What you need

READ THIS FIRST

To make the experiment as fair as possible DON'T read below the line, until you have written your FIRST thoughts in the spaces....

THINK OF

A colour
A wild animal
A flower
A piece of furniture
A fruit
The capital city of a country
A number from 1 to 10

WRITE IT HERE

AFTERWARDS

For each separate task (such as 'think of a colour') compare your thoughts with all the other children in your class. Your teacher will help.

You could display results like this :-

RED ✓
WHITE ✓✓
BLUE ✓✓

Do you find that the choices vary a lot? (Many _different_ responses.)

Or do large numbers of people have the same thoughts?

Use the knowledge you have learned from these tests to go into another class in your school — and pretend to be a mind reader.
(You could claim you will be able to tell what MOST people in the class are thinking ....)

Who else might be interested in knowing the ways people are most likely to think and choose?

© Collins Educational Science Teachers

You may photocopy this page for use within the classroom.

Reflections on Lucy...

What you need

Mirror (a plastic one is safest)
Pencil and crayons

Instructions

By holding a mirror vertically across Looking Glass Lucy you can make a great variety of symmetrical patterns.

Half of each pattern will be on the paper and the other half on the mirror.

I have provided eight examples of the sort of patterns you will see — but only FIVE of them can be made with the aid of a mirror.

DISCOVER WHICH PATTERNS ARE IMPOSSIBLE.

TRY TO GUESS FIRST — before using the mirror on Looking Glass Lucy.

Colour Looking Glass Lucy first, then colour the five **possible** patterns.

Discover some other different symmetrical designs and draw them.

© Collins Educational Science Teachers

You may photocopy this page for use within the classroom.

What you need
Writing and art materials

INVENT A TOY FOR A CHILD

Decide how old the child will be.
Will it matter if the child is a boy or a girl?

Try to think of something entirely different.

AIM FOR SIMPLICITY.

The toy must be easy to make......or manufacture.

Backroom in the Toy Business.

"Very good, Maddocks, but is there a demand for a ball that doesn't bounce..!!?"

Perhaps the toy will be intended to comfort the child.
Toys that give children something interesting to do are excellent.
A toy might be used to make learning fun.

You might be able to devise a new and interesting puzzle.
Colour, shape, size and decoration are important.
What about safety?
Noisy toys are interesting too!

Make a clear picture of your invention and write a description of it in the form of an advertisement.

It would be great if you could actually make and test the toy.

ELECTRIFYING Magic

What you need
Straight plastic straws
Scissors
Cotton
Adhesive tape

Try to get a straw to stick to your hand. Hold the straw in one hand, while stroking it with your other hand. USE DRY FINGERS.

After stroking, the straw should stick to your palm — you should be able to get it to stick under your hand. This stunt looks like a magic trick!

Stroking charges the straw with static electricity — that stays in one place and does not flow like electric current.

Get straws to stick to vertical surfaces, such as walls and windows. Try putting them on the ceiling. How long will they remain in these places, without falling?
Try to do these tests in both damp and dry conditions....

What length is the shortest straw you can 'stick up' in this way?

Dangle a charged straw from a thread, using tape, like this :-

What do you see when you move another charged straw near the dangling one?

Can you drive the dangling straw to swing in a circle?

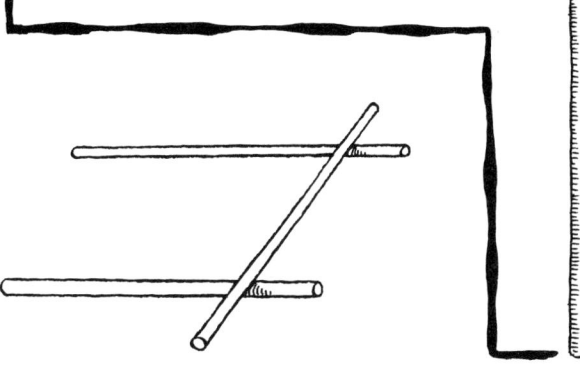

Rest a charged straw across a track made by putting two uncharged straws on the table. What happens when you hold a second straw close alongside the one on the track?

INVENT DIFFERENT THINGS TO DO WITH ELECTRIFIED STRAWS.

© Collins Educational Science Teachers

Little Dramas with Surface Tension

What you need

Washing-up liquid
Cocktail sticks   Drinking glass
Handful of small coins
Soup plate or shallow bowl

The surface of the water acts like a skin. This is not the same as saying there is a skin on the water. The water just acts as if there is, although you need to do some tests to prove it.

This imaginary 'skin' is called surface tension. If you were a tiny, newly-formed frog, trying to get out of a pond, surface tension would be quite a fence for you to break through.....

Test 1 Fill the glass to the brim with water. This glass should first be washed with washing-up liquid, and then be well rinsed with clean water.

How many coins can you insert through the surface of the water, without the water overflowing? Surface tension lets the water rise above the rim of the glass.

Test 2 Detergent weakens surface tension.
Float 3 clean cocktail sticks on clean water inside the clean soup plate. Use a fourth stick to get the others to float in the middle of the plate × as a triangular raft.

Dip the fourth stick in washing-up liquid, and then touch the water in the middle of the raft with the detergent-wetted stick. The three sticks float apart, as if by magic!

They are pulled apart by a stronger surface tension outside the triangle. Detergent has weakened the tension inside.

CLEAN EVERYTHING BEFORE YOU REPEAT THE TESTS.

Incredibly Musical Straws

What you need

Paper straws
Scissors
Sellotape

Squeeze one end of a paper straw. Snip off the corners of the flattened part.

Put this end of the straw in your mouth.
BLOW HARD. Don't make the straw too wet.
You should produce a pleasant musical tone.

THIS IS A SKILL - SO PRACTISE!!!

Musical pitch is highness or lowness on a scale. Pitch must not be confused with loudness or softness.
How is the pitch of a tone affected by making a longer pipe? (Straws can be taped together.)
How is the pitch affected by cutting a straw shorter?

Straw Pan Pipes

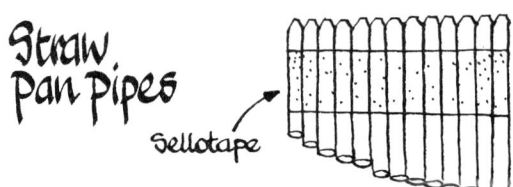
Sellotape

Cut a hole in a straw pipe. Cover and uncover the hole, as you blow....
Convert a straw pipe into a louder sounding "trumpet".
Snip pieces off the end of a pipe. As you do this, play a musical scale. (Perfect this skill as your "party piece"!)

Use straw pipes to imitate sounds of, for example, a police siren, a bassoon or the "cow in agony" that Mood Indigo!

What musical instruments depend on the scientific ideas that can be learned from musical straws?

CREATIVE BALLOONING

What you need

Materials for drawing and colouring pictures.

DRAW YOUR INVENTION OF A NEW USE FOR A BALLOON

Spend time thinking and doodling, before you draw your idea.

It can be a party balloon or a giant-size balloon.

The invention can be funny or serious. KEEP IT SIMPLE.

Try to invent an idea that you might actually make up and test.

Make a neat coloured picture. You can add a few words, if you wish to.

OR — draw a funny cartoon about a balloon or balloons.

LEAVE TIME TO SHARE YOUR IDEAS

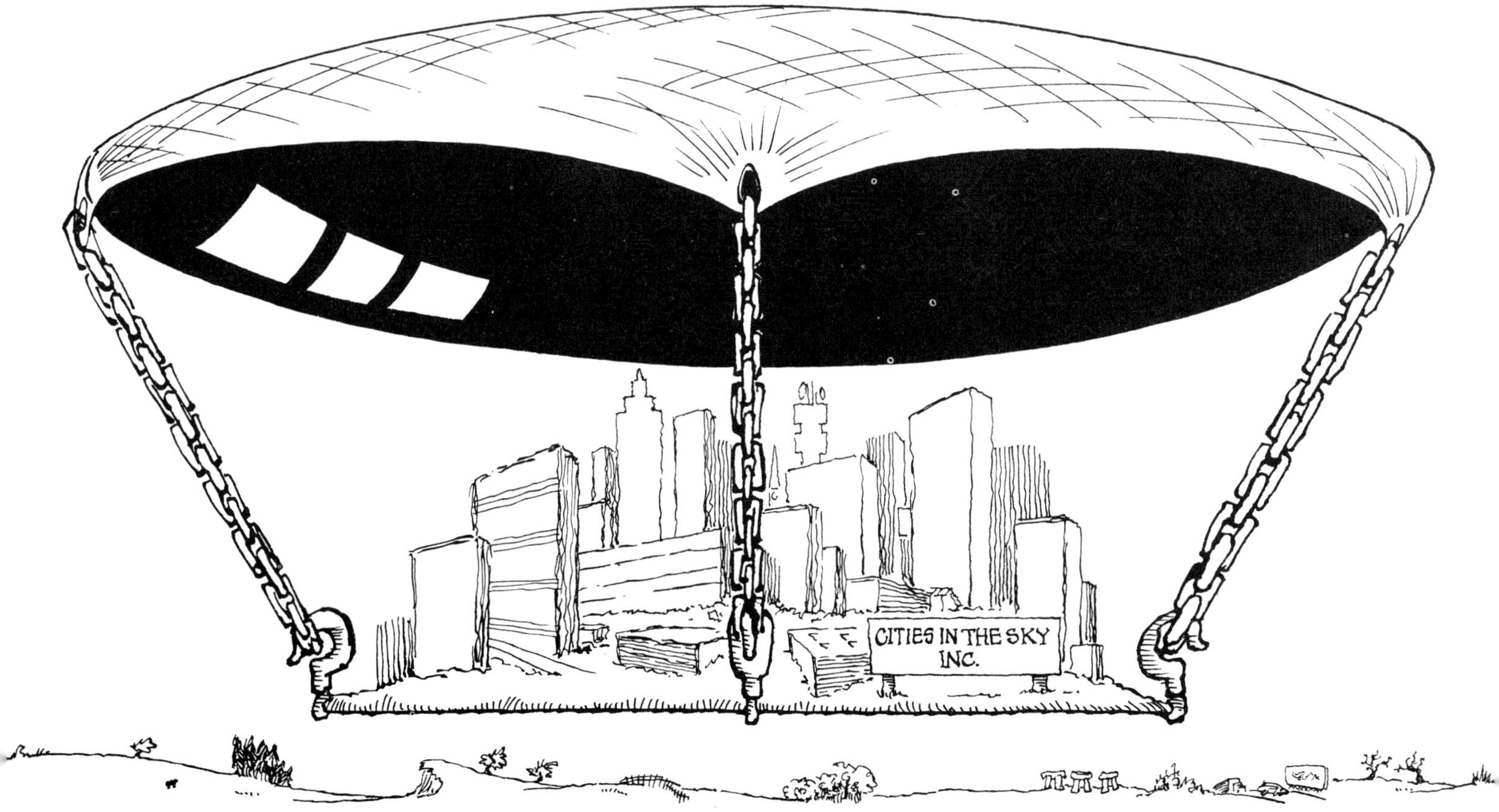

CITIES IN THE SKY INC.

© Collins Educational Science Teachers

You may photocopy this page for use within the classroom.

What you need

It depends which items you choose

Both the scientist and the artist need to be good observers using all their senses.

Here are some opportunities to put into practice your own skills of observation.

Aim for a score of 10, but make at least five different observations about one or more of the following:-
- Lighted candle - standing safely in sand
- Ordinary plain postage-stamp
- Small coin
- Queen of Hearts (playing card)
- Buttercup or other common flower
- Old teddy-bear
- Optical illusions in an aquarium
- Ice cube
- Grains of sugar
- Your left hand
- An outside wall

In the observatory...

The good observer also creates.

Choose the most interesting object you have observed and write a poem or short paragraph about it here:

....or you might prefer to demonstrate your skills as an observant artist instead.

© Collins Educational Science Teachers   You may photocopy this page for use within the classroom.

HOW MUCH AIR CAN YOUR LUNGS HOLD?

What you need

Large glass sweet jar.
Bowl or sink in which to submerge the jar.
Ping pong ball.
About a metre of thick plastic tubing.
A ruler

Completely submerge the jar in water in the bowl, letting the water replace all the air inside the jar. Stand the water-filled jar up side-down (water does not fall out....).

Put the ball up under the jar, to show the water level.

Poke an end of the tubing under the rim of the jar.

To find out how much air your lungs can hold, take a deep breath - then blow all this air through the tube, to collect the air under the jar.

Measure the depth of air collected.

Test each person in your group or class.

Make a graph like this:

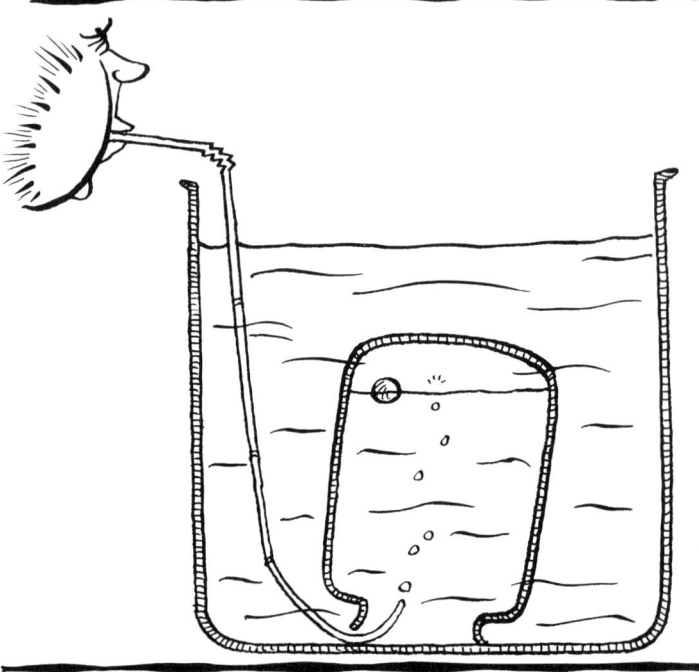

Is it important to practise before doing a "TEST BLOW"?

Can girls' lungs generally hold more air than boys' lungs?

Measure the amounts of air in cubic centimetres.

Represent your results on a different sort of graph.

What conclusions can you come to after doing the tests? (Note the physical dimensions and fitness of the children.)

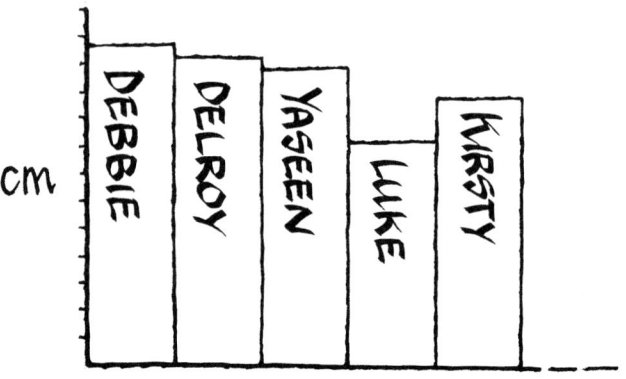

NOTE: A GIANT JAR THAT HELD PICKLES WILL BE MORE SUITABLE FOR USE WITH OLDER CHILDREN.

© Collins Educational Science Teachers   You may photocopy this page for use within the classroom.

Try your strength

What you need

2 kg bag of potatoes or a 2kg weight
Clothes pegs
Sweets
Timers

Test 1
Do push-ups with a bag of potatoes.... A 2kg bag should be about right ~ but feel free to change this. Everybody tested must use the same weight.

Keep pushing the bag up from shoulder height. How many times can you do this in one minute? Which of your arms is the most powerful? Who is the potato push-up champion in your class?

Test 2
How many times can you squeeze open and shut a spring-type clothes peg (sometimes called a clothes pin) in a minute? Repeat the test after one minute's rest.

Which of your hands is best at doing this task? Who is the champion squeezer in your class?

Test 3
How long can you hold your arm outstretched ~ at the height of your shoulder? Compare each arm ~ and compare your times with the other children's results.

I have heard somewhere that, after eating a sweet, you can keep your arm up longer. Enjoy a sweet while you rest for 10 minutes. Test your arms again. Is there any evidence that eating the sweet improves your performance?

DISCUSS WAYS TO MAKE YOUR TESTS FAIR. INVENT A NEW TYPE OF STRENGTH TEST.
After the sweet experiment, find out and talk about placebos.

DON'T EVER STRAIN YOURSELF!

What you need

Woodlice (search under a stone or rotten wood outside)
Hand-lens
Drawing materials
WOODLICE CAN BE HANDLED GENTLY, USING A PAINT BRUSH AND PLASTIC SPOON.

WHERE DID YOU FIND YOUR WOODLICE?
Was the place dark or light? Dry or damp? Did you find any small ('BABY') woodlice?
Are all your wood lice of the same species? Did you find a species that rolls into a ball? (AMERICANS CALL THIS A PILL-LOUSE)

Look at a woodlouse, using the hand-lens.
Make a careful drawing of it.
How many legs? Divisions on its body?
How much bigger is your drawing than the real woodlouse you copy?
Draw its front end, with correct feelers and eyes. Copy its back end accurately.

granfer Kruger... THE WOODLOUSE

How is your woodlouse like and unlike an insect or worm?
Do a piece of writing entitled:- "The Woodlouse Races"

Invent another name for a woodlouse. Vote on the most popular name invented by your friends.

Granfer Kruger (or Grampi Krugis) is a name used by some country people.

© Collins Educational Science Teachers

You may photocopy this page for use within the classroom.

"THE ACT OF RISING BY VIRTUE OF LIGHTNESS"
Chambers Dictionary

LEVITATION

What you need

Groups of six (similar body build and weight)
Plenty of space, with a few chairs
Determination to succeed

Person number one sits, arms folded, in a chair. One person (Number Six) acts as leader. HERE ARE THE INSTRUCTIONS FOR THE OTHER FOUR PEOPLE:

Person Two – puts both first fingers under the sitter's left armpit.
Person Three – puts these fingers under the sitter's right armpit.
Person Four – puts these fingers under the sitter's left knee.
Person Five – puts both first fingers under the sitter's right knee.

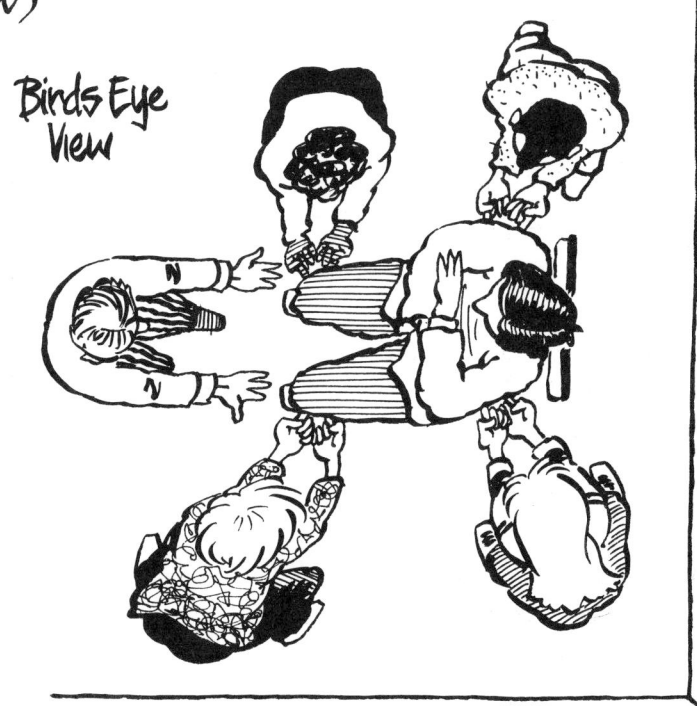

Birds Eye View

[ONLY THE FIRST OR FOREFINGERS OF THE LIFTING TEAM ARE TO BE USED]

The leader simply tells the team to try and raise the sitter into the air.

Working in their separate ways, the members of the team will almost certainly fail to raise the sitter effectively.... They need to act together in a co-ordinated way ~ they need to be able to exert their forces at the same instant.

THEY NEED A LEADER WHO WILL INSIST THEY 'GET THEIR ACT TOGETHER'.

SO TRY AGAIN... The sitter relaxes. Team members stand or kneel comfortably and extend their paired forefingers and insert them in the appointed places (in armpits, under knees).

The leader says: 'Ready ~ Steady ~ GO!!'
THIS TIME THE TEAM SHOULD SUCCEED BRILLIANTLY... but practise!!

| IF YOU HAVE SUCCESS WITH THIS EXPERIMENT, THE MOST PROBABLE LESSON YOU WILL LEARN IS THE POWER OF A TEAM ACTING IN A DISCIPLINED WAY. TRY IT....
What do you think?

© Collins Educational Science Teachers

You may photocopy this page for use within the classroom.

POSTCARD GIRDERS

What you need

Used A4 size papers
Cardboard cut to postcard size
Paperback books – they should be equal in weight

Take a piece of A4 paper. Fold the paper long ways, in 'accordian pleats' – like this……

– to make an object looking like the folds of a fan, or corrugated iron.

Put this paper, to form a bridge, spanning a gap between your friend's open hands (held about 20 cm apart).

You should be able to balance a paperback book on this model bridge, without the paper giving way.

SEE HOW STRONG PAPER CAN BE – IF IT IS WISELY USED!!

NOW TRY THIS: By bending a postcard, you can make it stand up on end – that's how you stand up a birthday card on a shelf.

By bending a postcard in different ways it can be made into variously shaped 'girders' – on which you can balance several paperback books.

Who can make a postcard girder to support the weight of most books?

© Collins Educational Science Teachers   You may photocopy this page for use within the classroom.

Cut the paper into halves — each will measure 7.5cm x 20cm.
Each piece will weigh the same.
With each piece of paper, make a different spinner.

Test the spinners by dropping them. They may need a bigger, smaller or an extra clip.
Which design works best? YOU MUST DECIDE WHAT "best" MEANS.

Whose spinner takes longest to fall 3 metres?
How will you make the competition fair?
Whose spinner has the most attractive decorations?

What you need

STIFF 15cm x 20cm paper
Metal paper clips — various sizes
Scissors
Electronic watch sensitive to 1/100 ths of a second

SPINNER X

Cut the paper like this.
Bend up the wings.
Put a clip on the bottom end.

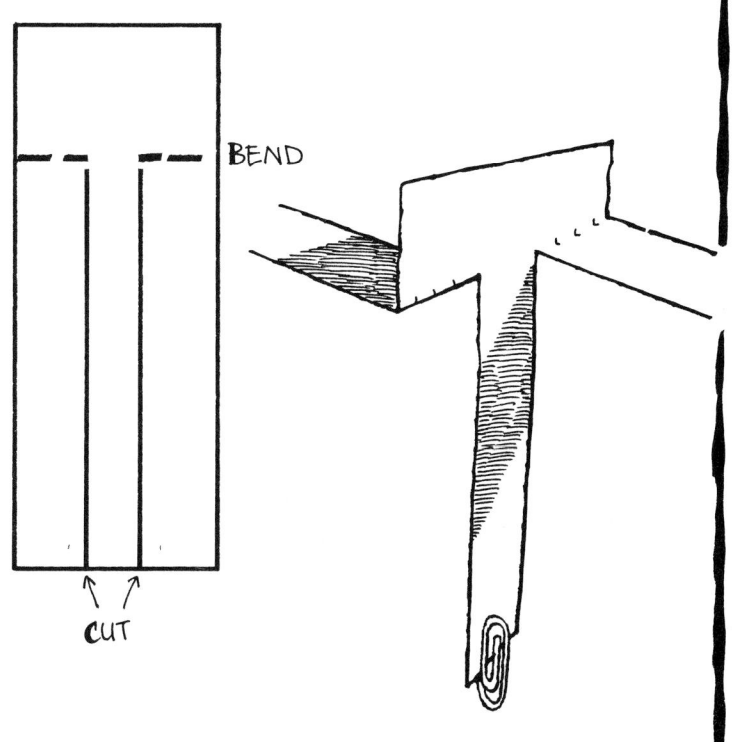

SPINNER Y

Cut the paper like this.
Bend up the sections of the bottom end, to make several thicknesses of paper there.
Bend out the wings.
Put a clip on the bottom end.

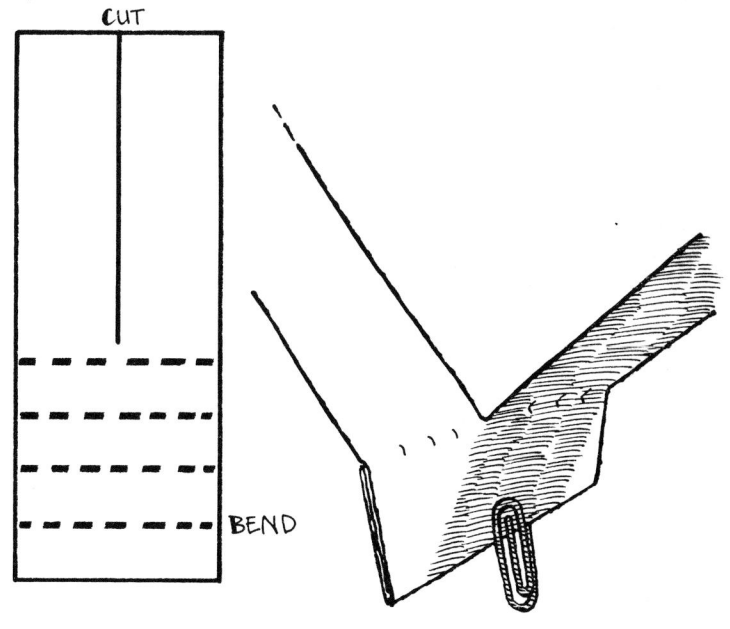

© Collins Educational Science Teachers

What you need

Time to talk about the correct answers.
Access to reference books, if possible.

Put a CROSS next to the name of each imaginary animal in the list.

Underline the names of extinct animals – those which once upon a time lived on the earth, but are now all dead.

Draw a ring around the name of every animal still alive today.

THE ULTIMATE ZOO QUEST

Passenger Pigeon
Toucan
Phoenix
Duck-billed Platypus
Pterodactyl
Flying Fox
Sabre-toothed Tiger

Wombat
Centaur
Dodo
Sloth
Mermaid
Stegosaur
Coelacanth

Kraken
Unicorn
Narwhal
Mammoth
Bunyip
Griffin
Great Auk

Inside this enclosure, draw your impression of one animal from each group:

Extinct   Living   Imaginary

Look in reference books to find out about the animals you get wrong.

© Collins Educational Science Teachers   You may photocopy this page for use within the classroom.

Design a New Kind of Insect

What you need

Textbook references about insects (optional)

THE 'TYPICAL' INSECT HAS:-
- HEAD, with eyes, mouth parts and feelers.
- THORAX (middle part), with two pairs of wings and three pairs of legs.
- ABDOMEN (rear part, often quite long), which in the female may have a sting or egg-laying tube.

Mouth parts can be ingenious, depending upon how the insect feeds. They might be specially shaped for chewing, or for piercing and sucking.

One of the pairs of wings might be used as a tough covering for the other pairs of wings, when they are folded underneath — as with beetles. Or the wings might be missing altogether — as with worker ants.

DELROY WASHINGTON TAKES PET COCKROACH "BERTY" FOR A STROLL.

Play at being Mother Nature, **INVENT AN INSECT**. Make an attractive, large and detailed picture of your insect. Think of an apt name for it. Write a sentence or two about how it feeds, where it prefers to live and how it defends itself.

BUT... your idea only qualifies if its structure obeys the rules (see above). A scientist must be able to look at your picture and say: "That's an insect".

This game can be adapted to other animal groups ~ and plants....

© Collins Educational Science Teachers

Looking-Glass Races

PLACE EDGE OF MIRROR HERE ↑

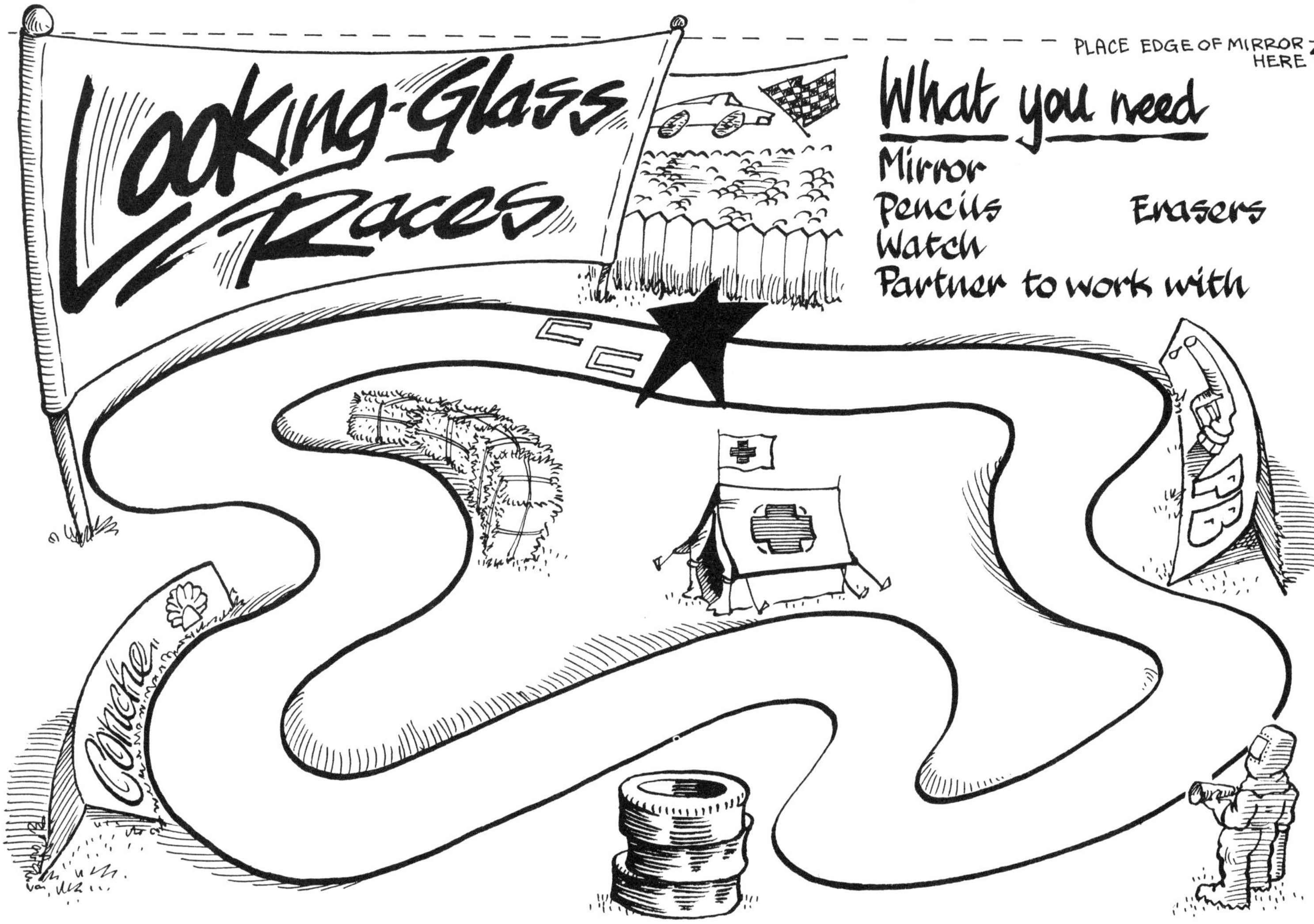

What you need
- Mirror
- Pencils
- Erasers
- Watch
- Partner to work with

FOLD UP THIS END OF THE PAPER ALONG THE DOTTED LINE.

Let this folded up part hide the real track from your eyes. Get your partner to hold the mirror on the far edge of the paper. You see the mirror-image of the track.

With your free hand, trace a pencil line around the track, starting and finishing at the star. But only look at the REFLECTION of your pencil and paper.

You have to avoid going over the borderlines of the track.

You may be reminded of what you felt like when you were being taught to write. Following the reflection means having to learn different ways of using your muscles.

Rub out your first pencil line. Is following the track easier the next time you try?

Let your partner do the timing, while you make three timed circuits of the track. Do you get quicker with practice?

Do the timing while your partner has some turns.

Can anybody in your class complete the circuit without pencilling over the borderlines?

© Collins Educational Science Teachers

You may photocopy this page for use within the classroom.

Observe & Remember

A game of skill for two or more players.

What you need

Several packs of cheap playing cards (Snap picture cards provide an easier game)

PLAY THE GAME ON TABLES WITH LARGE SURFACES, NOT DESKS.

Shuffle the cards. Spread them all over the table, by putting down each card separately, with its face hidden. You can only see the back of the cards.

THE OBJECT OF THE GAME IS TO WIN THE MOST PAIRS OF CARDS.

A pair consists of any two cards that have the same face value — for example: two Aces, two sevens, or two Queens.

The first player turns up two cards – to show their faces. If they happen to match they are a pair. The player wins them and has another go.
If the cards do not match they must be turned down again, to hide their faces – and the next player has a go.

A SKILFUL OR LUCKY PLAYER CAN GO ON WINNING PAIRS, UNTIL TWO NON-MATCHING CARDS ARE TURNED UP.

It is important to observe the exact position of the cards you have seen and to remember their values. Always turn over a card you are not sure about first (you will soon learn why this hint is useful).

The winner of the game is the player who has the most pairs when all the cards are used up.

© Collins Educational Science Teachers

You may photocopy this page for use within the classroom.

dancing grapes

What you need

Drinking glasses or jam-jars
Large bottle of fizzy lemonade
Grapes

Pour a glass of lemonade. Put in about five grapes. Watch the grapes rocking and rolling amongst the bubbles.

The gas in the bubbles is carbon dioxide.
Where does the gas come from?

What do you think prevented much fizzing in the unopened bottle?

Grapes normally sink in lemonade. What is keeping them afloat?
   You could time how long it takes for the first grape to sink.
   Why do the grapes start to sink?
   You could time how long it takes for the first grape to sink, to float to the surface again.

EVENTUALLY THE GRAPES ARE ALL "DANCING" UP AND DOWN.....

Write a poem ~ or a paragraph ~ describing what you have observed.

Do other objects behave in lemonade like a grape? (Try mothballs and peanuts.)

When are grapes: denser than water?
less dense than water?
most buoyant?
less buoyant?

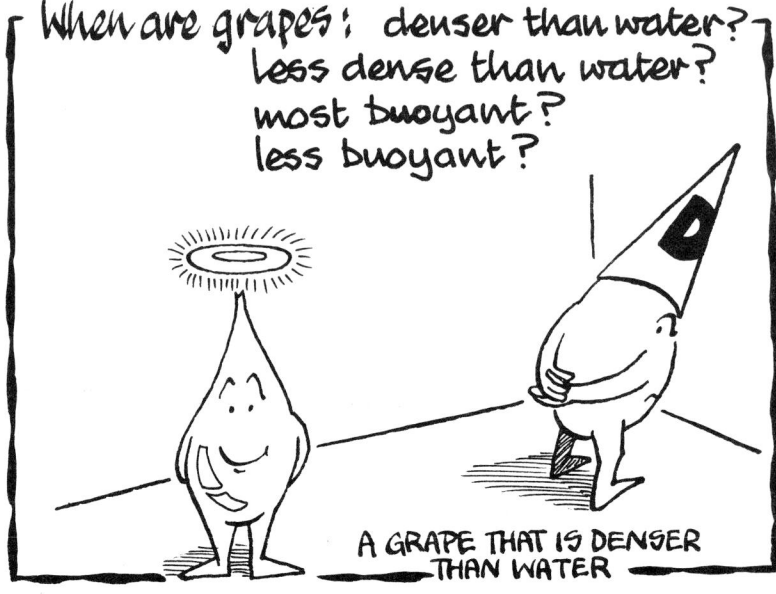

A GRAPE THAT IS DENSER THAN WATER

© Collins Educational Science Teachers

You may photocopy this page for use within the classroom.

Everybody's Hang Glider

What you need

A4 size papers of different weight
Box of paper clips.

PLENTY OF SPACE INDOORS

Fold a sheet of A4 paper in half across its narrowest part (its width). Clip both thicknesses together, near the fold — by putting a paper clip at each end of the fold. The clips go in line with the fold.

Curl the halves of the paper outwards and downwards. Then clip together the corners A and B — which will be at the front end of the hang-glider.

The structure is very floppy and flexible. It is a kind of non-rigid 'Rogallo Wing', named after its inventor — although this clever paper model was devised by Johnny Ball.

Form a short chain of clips, attached to the clip at the front end of your hang-glider

Test the model by trying to make it glide in a steady way, but not losing height too soon. (You can vary the number of clips in the chain.)

Research and Development

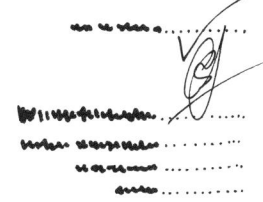

- Use different weights of paper.
- Change the position of the clips.
- Use clips of a different size.
- Design and test a bigger glider.
- Cut slots and holes in the paper.
- Vary the shape of the paper.
- Colour your glider — attach a 'pilot'.
- Test your model outdoors.

Whose glider has the longest flight-path, when dropped from a height of, say 1.5 metres?

© Collins Educational Science Teachers

What you need

Scraps from cardboard boxes (the paler the better)
Colouring sticks or pencils
Scissors

Design your animal in such a way that by far the greater part of its cardboard mass is going to be underneath the place on its shape that is intended to balance on a pencil.

HERE ARE TWO SHAPES YOU COULD TRY:-

JOIN THE BALANCING CIRCUS

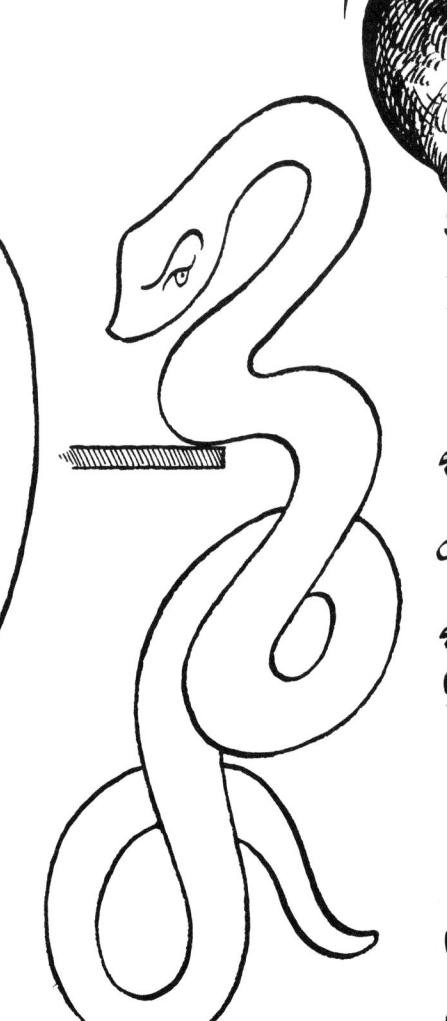

Either Or

Two Possible Perches

Begin by thinking hard about the shape of your creature.
Draw its outline on cardboard - and cut it out.
Test your idea by trying to get the shape to balance.... Then decorate your SUCCESSFUL balancer. Use bright colours.
(Make your idea as funny as you wish.)

MAKE A BALANCING MOON GIRL OR A FUTURISTIC ROBOT.
Make a cheerful display - call it the "BALANCING CIRCUS."

© Collins Educational Science Teachers   You may photocopy this page for use within the classroom.

The art of being hidden in full view of your enemies

What you need
Crayons and painting sticks
Paste
Scissors

Draw and colour an animal (mammal, fish or bird).
It may have spots or stripes.
You could invent an animal.

YOUR ANIMAL HAS ENEMIES.

Draw and colour a background (grass-land, deep-sea, jungle) where it would be difficult for an enemy to see your creature.

Cut out the animal and paste it on the background.
How well is the creature camouflaged?

Plan and do a test to decide which child in your class has the best camouflaged animal.

© Collins Educational Science Teachers

Would you have survived a Wild West Gunfight!!?

What you need
Rulers calibrated in centimetres (cm)

REACTION TIME is how long it takes from when you sense danger to when you have acted to protect yourself from harm. Here is a rough test of your reaction time. It is measured in decimal parts (**fractions**) of a second.

TEST
Let your friend dangle a ruler, with its zero mark exactly in the middle between your forefinger and thumb – which should be held open like claws, 3 cm apart.

Tell your friend to let go of the ruler – but your friend must not tell you exactly when.... As soon as you sense the ruler is falling, catch it, grip it – DON'T LET IT SLIP THROUGH YOUR FINGERS.

Read the measurement on the ruler, where you have caught it. Look for the measurement on the chart of reaction times. The chart shows how long it took from the moment you saw the ruler falling, until the instant you caught it – THIS WILL BE YOUR REACTION TIME.

DOES YOUR REACTION TIME IMPROVE WITH PRACTICE?

Is your reaction time slower when you are tired (say after a long busy day at school)?

Chart of Reaction Times

Distance of fall (cm)	Time of fall (sec)
1	0.045
2	0.064
3	0.078
4	0.090
5	0.101
6	0.110
7	0.120
8	0.128
9	0.136
10	0.143
11	0.150
12	0.157
13	0.163
14	0.169
15	0.175
16	0.181
17	0.186
18	0.192
19	0.197
20	0.202
21	0.207
22	0.212
23	0.217
24	0.221
25	0.226
26	0.230
27	0.235
28	0.239
29	0.243
30	0.247

Imagine and write down five instances where a quick reaction time could save you from danger. Make a list of jobs for which having quick reaction times would be advantageous.

© Collins Educational Science Teachers

You may photocopy this page for use within the classroom.

LIQUID engineering
BLOWING BUBBLES

What you need

Washing-up liquid
Cups or yogurt pots
Plastic drinking straws
Scissors
Water — TAKE CARE NOT TO DO ANY DAMAGE!

The bubble-making mixture is prepared by dissolving about a teaspoonful of washing-up liquid detergent in half a cupful of water.

THE STRAW

Make a bubble-pipe from a plastic drinking straw. Squeeze one end of the straw, to flatten it. Cut two slits along the flattened part, using scissors.

Bend out portions of the straw between the slits, to make an object resembling a small flower. The bubbles will be formed at this end.

Dip the cut end of the straw in the mixture. A drop of liquid will be trapped inside the pipe.
PRACTISE BLOWING BUBBLES
Hold the pipe away from the cup, with its wet end slightly downwards — and blow slowly, gently.....

WHO CAN BLOW THE BIGGEST BUBBLE? Whose bubble (the size of a cricket ball) lasts longest?

How many colours can you identify in a bubble?

Find surfaces on which bubbles will bounce.

Draw and colour an accurate picture of a bubble.

How is a bubble like (or different from) a water drop, or a rubber balloon?

© Collins Educational Science Teachers

DOMINO ENGINEERING

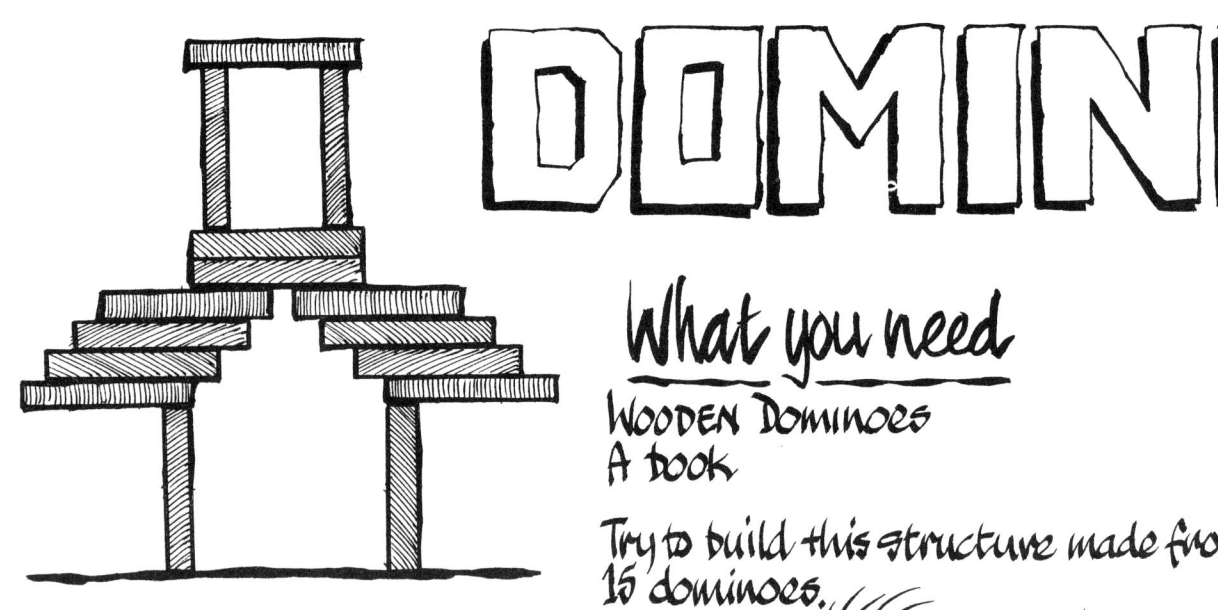

What you need
WOODEN DOMINOES
A book

Try to build this structure made from 15 dominoes.

It looks impossible!

HINT - perhaps you need a few extra dominoes...

Make a line of standing dominoes in such a way that, when the first domino is pushed, the others will be pushed over, one after the other. The dominoes are being made to act like a fuse, which can make a bomb explode.

This book is balanced on the edge - a slight push will cause it to crash on to the floor. Form a 'fuse' with dominoes, which - on being activated - will make the book fall over the edge.

It's fun to make a really long 'fuse'.....

I have just done this successfully, but I must admit I added a ruler to my equipment.

© Collins Educational Science Teachers

You may photocopy this page for use within the classroom.

SURVIVAL KITS.

What you need

It would be interesting to have an actual kit made up. A SOLDIER'S BASIC SURVIVAL KIT CAN BE PUT INTO A MATCHBOX. The box would contain:-

- waxed matches
- rubber balloon
- fish-hook
- large polythene bag
- chalk
- needle
- razor blade

Imagine you were a man or woman soldier in an emergency. Think about what sort of emergency it might be – DO THINK OF SOMETHING EXCITING! Then tell your story of how you used your matchbox survival kit. Think carefully about how the items might be used in really clever ways, before writing your story.

Aim to get your story written in time to share it with other imaginary adventures in your class. Write it simply, so it will be easy to read aloud.

INVENT A BASIC SURVIVAL KIT FOR LIFE IN SCHOOL......

HAHA HA! Oh! That's good!!

© Collins Educational Science Teachers

WONDERS IN THE SKY

What you need
Patience and imagination.

Close both fists. Poke out your forefingers (first fingers). Touch them together, as if they were pistols aimed at each other. Raise them to a position 20cm in front of your eyes — and on a level with your eyes.

Look past your fingers, at some distant place. While you do this still be aware — WITHOUT ACTUALLY FOCUSING YOUR EYES ON THEM — of your touching fingers.

You should see what looks like a small sausage. It seems to be supported between your fingertips.

Slowly pull the fingers a few centimetres apart. The "sausage" now seems to be floating in mid air. If you look at it against the sky (not against the sun), you get an impression of an unidentified flying object — a UFO.

Your fingers are out of focus, so your eyes see the fingertips 'double'. The out-of-focus images in your mind overlap, so you see the 'UFO'.

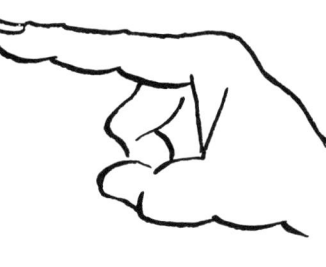

DO YOU BELIEVE IN UFOs? These objects in the sky look mysterious:
- LIGHT REFLECTED FROM CLOUDS
- CLOUDS SHAPED LIKE WHIRLPOOLS
- THE PLANET VENUS (the morning or evening 'star')
- SHOOTING-STARS (meteors)
- MAN-MADE SATELLITES

ADD SOME MORE ITEMS TO THIS LIST.

Advertisements have been projected onto clouds, using searchlights. List some other ways that advertisements have appeared, or might be made to appear in the sky. You can draw some if you wish......

———— If you have a theory about UFOs, tell it to your class. ————

© Collins Educational Science Teachers

You may photocopy this page for use within the classroom.

Who believes in Ghosts?

✱ Personally, I doubt if GHOSTS ARE real in the sense that they occur outside of human imagination. But I am in no doubt that the thought or idea of ghosts is real, in the sense that people can be scared by something in their minds. I do not believe in the supernatural—that would be against the laws of nature. I believe everything that happens, however mysterious, must have a natural explanation, even if scientists are not yet wise enough to know the explanation. Perhaps they never will be!!

Quite simply: DO YOU BELIEVE IN GHOSTS, YES OR NO?
(Make a graph for your class.) Do more BOYS in your class believe in ghosts than do GIRLS?

NOTE
IF YOU SUSPECT SOME CHILDREN MIGHT NOT WISH TO ADMIT IN PUBLIC THAT THEY BELIEVE IN GHOSTS, YOU COULD PLAN A SECRET WAY TO ANSWER THE QUESTION.

DISCUSSION
What sorts of events in a house might be mistaken for ghosts? (Changes in temperature have certain effects.)

What sorts of outdoor sights and sounds might be mistaken for ghosts?

When is a person most likely to be scared by the idea of ghosts?

If anything spooky has ever happened to you, write what happened, as accurately (but briefly) as possible. You could read out your story and ask your friends for their comments and questions.

Invent a plan to show how a magician or theatre director could arrange for a ghost to appear in a stage play.
(DRAW YOUR IDEA.)

ARE YOU A MIGHTY EGG-CRUSHER?

What you need

EGGS IN A CARDBOARD EGG-BOX.
PLASTIC BAGS.

Put an egg with its ENDS between the palms of your hands. Grasp it firmly.   DON'T TRY TO CRUSH IT ~ YET!!

Put your hands, with the egg, inside a plastic bag. Grasp the egg as you did before.

Now try hard to crush the egg between your hands.
(THE BAG WILL SAVE YOU FROM GETTING EGG ON YOUR CLOTHES)

Even if you are very strong, you will find it impossible to crush an ordinary hen's egg in this way. SPARE A MOMENT TO THINK IN WONDER. The double-dome shape makes a 'fragile' egg remarkably strong.   WRITE A LIST OF STRUCTURES STRENGTHENED BY BEING DOME SHAPED.

(What about a supermarket egg-box?)

My newspaper informs me that some children were UNABLE to smash an egg by throwing it high in the air ~ and letting it fall on grass.
It might be possible to go outside and do some tests.

HOW MUCH PRESSURE IS NEEDED TO CRUSH A SUPERMARKET EGG-BOX? (Plan and do a fair test. Test several boxes.)

Bathroom scales will help you measure this force.
You can use people and books as masses.....

© Collins Educational Science Teachers

You may photocopy this page for use within the classroom.

EVERYONE'S HEART-THROB...
Feeling your Pulse

What you need
Drinking straws
Timers

FEEL EACH SIDE OF YOUR NECK, JUST BELOW THE CHIN. You can feel beats, or throbbing. This is called your **PULSE**. You are feeling blood being pumped from your heart, through pipes (**ARTERIES**).

Work in pairs. Listen by putting your ear to your partner's chest. You should hear the heart beating.

Find a place — SOMEWHERE ON YOUR WRIST, JUST AT THE BASE OF THE THUMB — where you can see or feel a pulse beating. (Feel gently with the tips of your fingers.)

Rest a drinking straw across this place. Watch the straw closely. You should be able to get it to jerk in time with your pulse.

THE **PULSE** IS TAKEN BY A PARENT, DOCTOR, NURSE OR A SCIENTIST – TO FIND OUT HOW MANY TIMES IT BEATS IN ONE MINUTE.

YOUR TEACHER COULD TIME ONE MINUTE FOR THE WHOLE CLASS. OR YOU MIGHT PREFER TO USE YOUR OWN WATCH.

AFTER RELAXING FOR A FEW MINUTES, count how many pulse beats you have in one minute.

Make a graph to show the 'RESTING PULSE' of everyone in the class.

How would you work out the 'AVERAGE PULSE' for a child of your age?

Plan and do an experiment to see how exercise changes your pulse.

Your pulse is supposed to beat faster when you are asked to do a tricky sum in your head.... Plan and do tests to find out if there could be any truth in this idea.

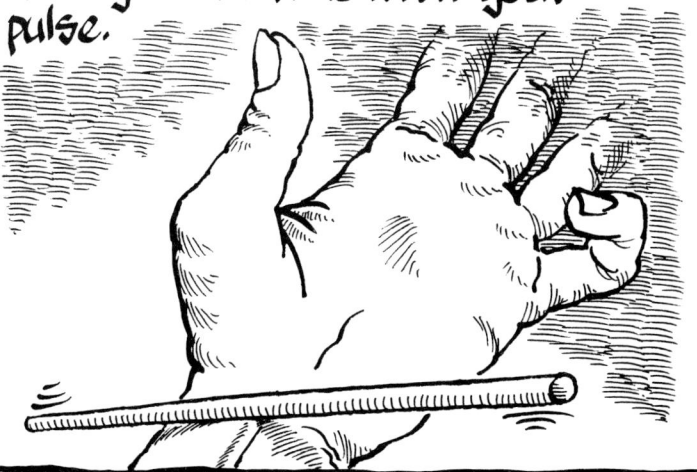

© Collins Educational Science Teachers

Draw a Healthy Person

I am sure that you have your own ideas about what a healthy person should be like.

Spend a little while thinking about this. If you like, you can make some rough notes.

Then make a big bold drawing of your ideal healthy child or adult. You can, if you wish, add extra information to the drawing by writing sentences and putting in arrows, to make your ideas clearer.

MORE IDEAS:
MY IDEAL TEACHER.
A TOTALLY UNHEALTHY PERSON.
AN IDEAL HUMAN BEING.

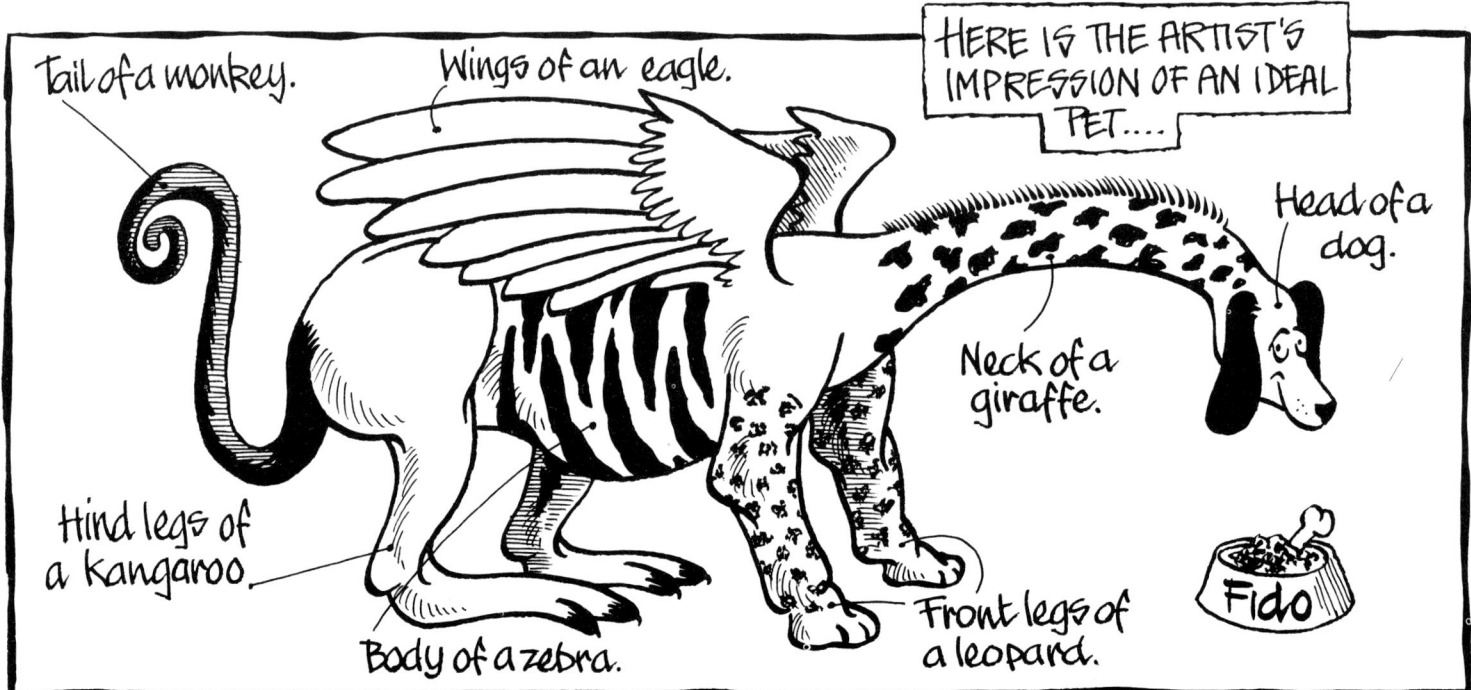

HERE IS THE ARTIST'S IMPRESSION OF AN IDEAL PET....
- Tail of a monkey.
- Wings of an eagle.
- Head of a dog.
- Neck of a giraffe.
- Hind legs of a kangaroo.
- Body of a zebra.
- Front legs of a leopard.

Could you talk to the class for one minute about your picture? Maybe you could also answer questions about your ideas....

© Collins Educational Science Teachers — You may photocopy this page for use within the classroom.

ELECTRIC GENYUS

What you need

Batteries, bulbs and wires (optional)

Assume that I am using ordinary 1.5 volt bulbs and 1.5 volt battery cells. They are connected with copper wires, bared at both ends.

WHICH BULBS WILL LIGHT?

WHICH WILL NOT LIGHT?

WHICH WILL LIGHT BRIGHTLY?

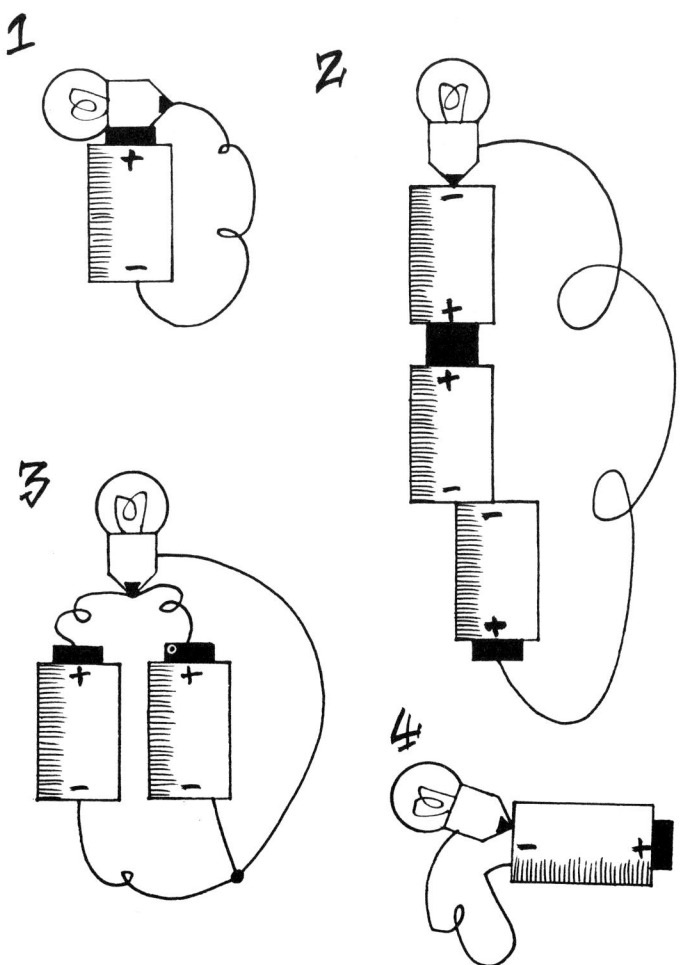

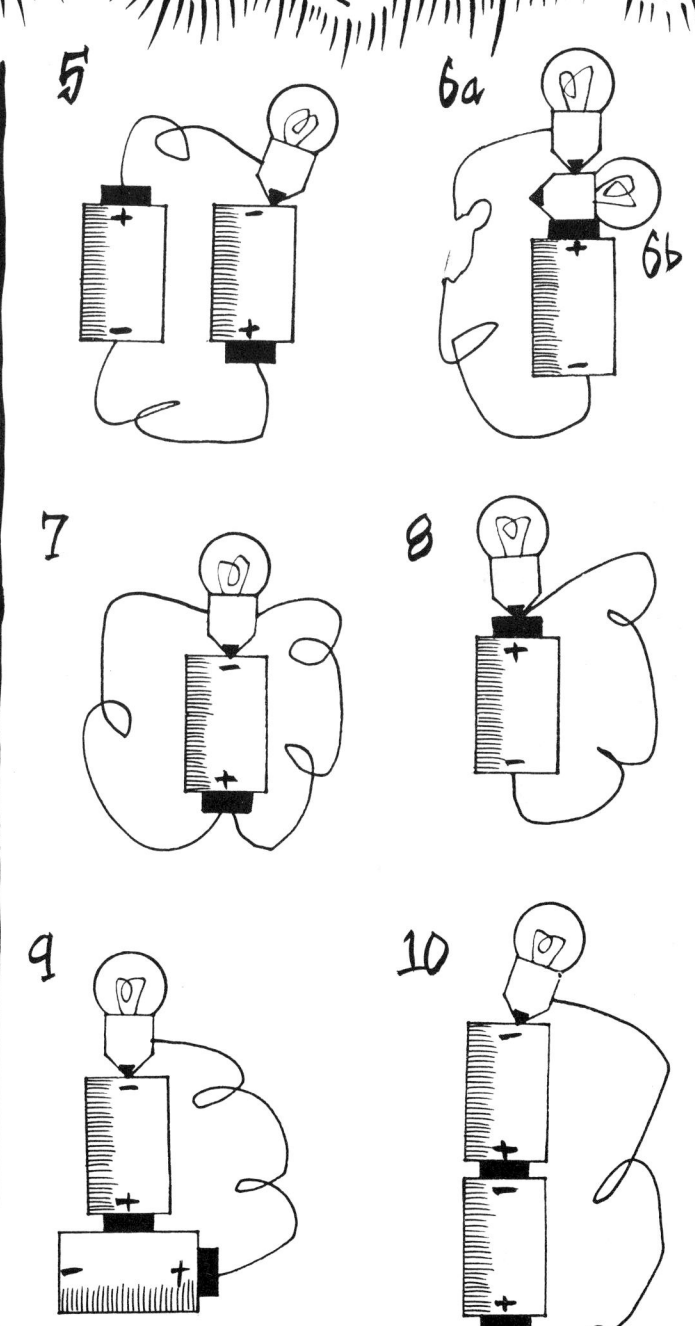

INVENT SOME MORE PUZZLING CIRCUITS.
HOW COULD THE BULBS THAT WILL NOT LIGHT BE MADE TO WORK?

© Collins Educational Science Teachers

You may photocopy this page for use within the classroom.

IDEAS IN A MILLION

What you need

A five pound note.
Ruler
Can of lemonade (displaying its contents in millilitres)
Newspapers
Calculators
THE GUINESS BOOK OF RECORDS might come in handy.

How many is a million? Yes, I know it is a thousand thousand. Have you ever tried to imagine a real million? If I told you that a new fighter aircraft for the RAF is going to cost taxpayers ten million pounds, have you any idea what that huge sum represents?

Measure the length of a five pound note. Try to work out how long the line would be if you put 2,000,000 of them end to end.

More ideas in a million

Find out whether you have lived for a million minutes.

Work out whether you could walk a million centimetres in one day.

What sort of animals would weigh about a million grams?

How many cans of lemonade provide a million millilitres of drink? (Base calculations on the contents of an actual can.)

Put together a newspaper containing a million cm² of paper. (Make the bundle of pages as accurate as possible.)

How long would it take you to type a million full stops? (Assume that you can type one every second.)

Invent your own "IDEAS IN A MILLION".

MORE liquid engineering

Extras Needed

Marbles
(I advise the use of formica topped tables)

On a table, wetted by spreading bubble-making mixture with your bare hand, you can create dome-shaped bubbles. Just hold the cut end of the pipe close to the wetted table ~ and blow bubbles, one at a time......

What pattern can be formed by joining bubble domes together?

Do you find more than three bubble surfaces joining along a single edge?

Can you estimate angles formed between the bubble domes?

Try bowling wet marbles through a bubble dome.

Who can put the most bubble domes !! INSIDE EACH OTHER !! like a nest of boxes? (You have to work out the most effective way to do this.)

Challenge

Who can make a bubble "caterpillar", like this:

METAL BOATS

What you need

Aluminium foil cut into squares (15cm × 15cm)
Marbles - they should each weigh the same.
Bowls or plastic boxes containing water.

Shape a foil square any way you wish, to make a boat.
Whose boat can carry the most marbles as cargo?
Which boat shapes seem to work best? (THEY MUST NOT CAPSIZE AND SINK!!)

Can you get your piece of aluminium to sink?
Which is denser, solid aluminium or water?

WHAT IS YOUR EXPLANATION for the fact that making aluminium foil into a boat makes it float well enough to carry cargo?

TRY OTHER CARGOES - for example, Lego squares of equal weight.
(Does your boat carry more or less Lego squares than marbles?)
PLAN A BOAT COMPETITION, USING PAPER INSTEAD OF ALUMINIUM FOIL.

ANIMATED GRASS HEADS

What you need
Bristly seed-heads of wall barley grass (HORDEUM MURINUM)
Paper tissues
Selection of feathers

Fold a paper tissue into a flat tube, 4 or 5 cm wide. Insert the grass head, point first, into one end of the tube. Grip the tube at each end — but avoid gripping the grass inside. Keep pulling (tightening) and pushing (slackening) the tube. This action keeps jerking the out-of-sight grass head.

The barley grass should travel along the tube and appear at the other end. **PRACTISE THIS.**

In a few words write your own explanation of how this scientific trick works. Compare your ideas with those of other children.

Who can make their grass head go along the tube in the shortest time? Who can do this with the least number of jerks?

--- **HOW WILL YOU MAKE SURE THAT THESE TESTS ARE FAIR?** ---

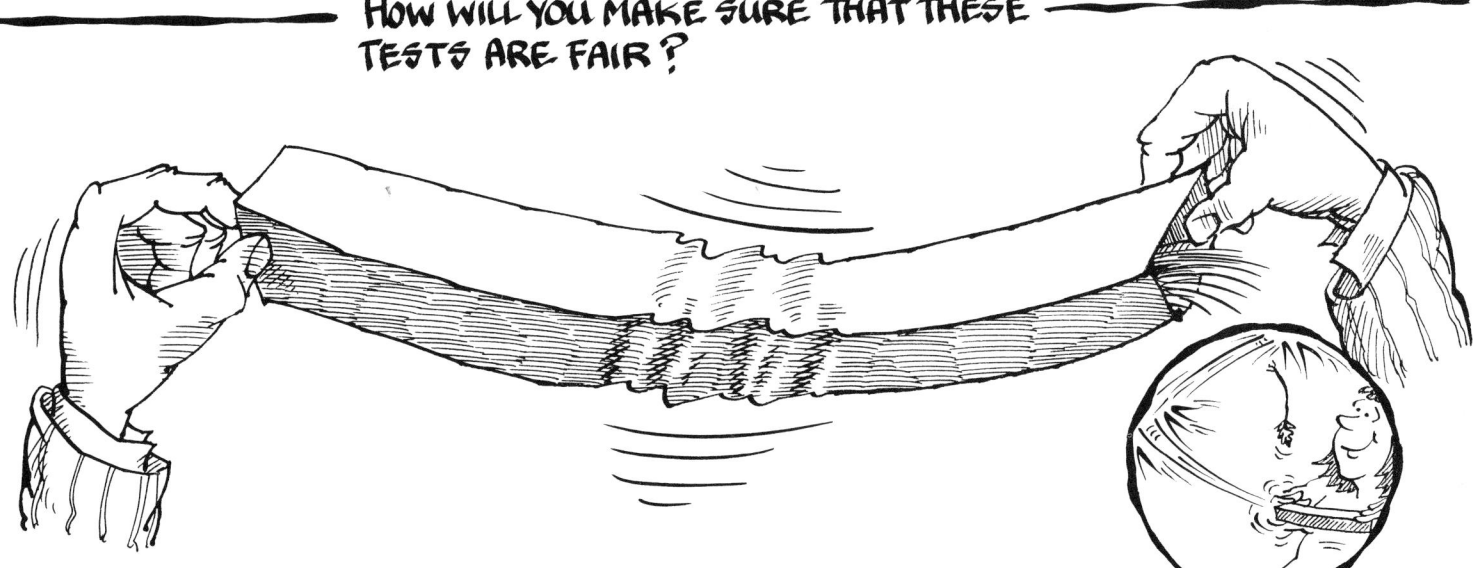

Hold races with say five children racing their barley grass per heat.

As a child I used to poke barley grass up inside my sleeve. After walking some distance I would find the grass inside tickling my armpit, after travelling up my arm.... DO TRY THIS.

Can anybody think of a use for 'barley grass technology'?
P.S. I HAVE JUST GOT THE TRICK TO WORK WITH A FEATHER......

© Collins Educational Science Teachers

Brainy Birds Competition

What you need

Moss, Sticks, Leaves, Scraps of wool, Bits of string, Shredded paper, Mud (optional) Etc.

IDEAL FOR A SCHOOL CAMP – OR FOR WORKING OUTSIDE ON WASTE GROUND

Talk about how different birds construct their nests.....

The Woodpigeon builds a platform of thin sticks.

The Magpie uses sticks to build a domed nest.

The Wren's domed nest is made of leaves.

The Housemartin uses mud to build a nest under the eaves of a house.

A Sparrow makes an untidy nest from dried grass and feathers.

Feathers, moss, even dried mud are used for nest linings.

Who can put together the most interesting model bird's nest?
Who can put together the neatest model nest?
Only natural materials to be used!
— USE ONLY YOUR HANDS —

A "BOOBY PRIZE" to the "BOOBY BIRD" for the most untidy nest....

FIRST HE SWALLOWED THE MUD.... THEN THE MOSS .... THEN LEAVES.... THEN...

Conservationist David Bellamy insisted on holding the materials with his mouth – I admire his dedication to realism, but please don't YOU risk swallowing something nasty!!

YOU WILL APPRECIATE THE SKILLS BIRDS HAVE.

© Collins Educational Science Teachers   You may photocopy this page for use within the classroom.

INSTANT CHALLENGES

What you need

Time to think

Count the squares, including overlapping ones, that appear in this network.

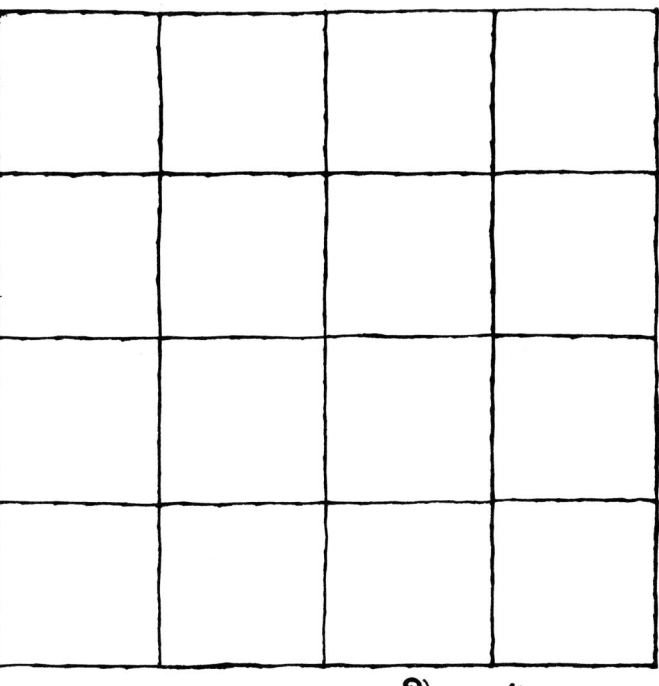

Total............

P.S. What is the total number of <u>rectangles</u> inside the network of squares?

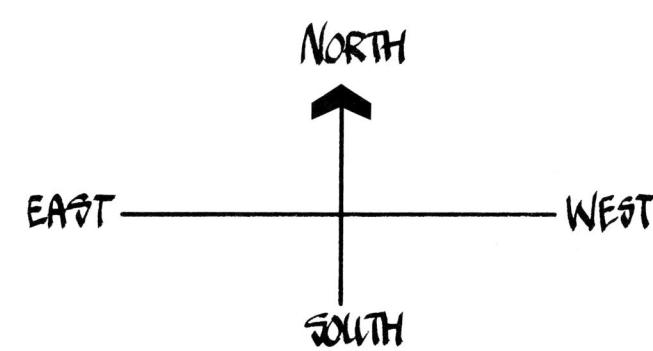

Where in the building in which you are working could you put this cross, to show correctly the points of the compass?

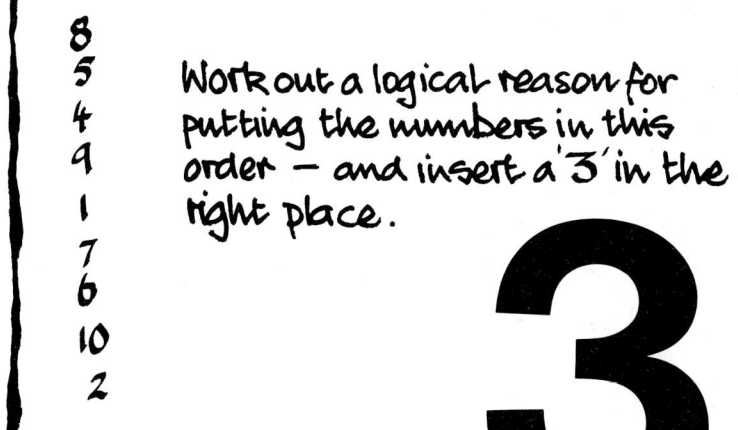

Work out a logical reason for putting the numbers in this order — and insert a '3' in the right place.

© Collins Educational Science Teachers

You may photocopy this page for use within the classroom.

Some Answers and a Hint

Cupid's Mysterious Love Letter.

Art thou not dear unto my heart?
Oh I search that heart and see
And from my bosom tear the part
That beats not true to thee.

But to my bosom thou art dear,
More dear than words can tell.
And if a fault be cherished there,
'Tis loving thee too well.

Reflections on Lucy.

Patterns 3, 5 and 8 are impossible.

The Ultimate Zoo Quest.

EXTINCT: passenger-pigeon, pterodactyl, sabre-toothed tiger, dodo, stegosaur, mammoth, great auk.

LIVING: toucan, duck-billed platypus, flying-fox (fruit-bat), wombat, sloth, coelacanth ('living fossil' fish), narwhal (whale).

IMAGINARY: phoenix, centaur, mermaid, kraken, unicorn, bunyip, griffin.

Domino Engineering.

A HINT
Use extra dominoes to provide support while you build the main and heaviest part of the structure - then remove the temporary supports.

Electrical Genius

WILL LIGHT: 1, 2, 3, 5, 6a, 7, 10.
WON'T LIGHT: 4, 6b, 8, 9.
EXTRA BRIGHT: 5, 10 (new cells will probably 'blow' the bulbs!)

Instant Challenges

30 SQUARES and 100 RECTANGLES.
Put the cross on the ceiling!
Insert 3 below 10 and above 2 - they are in alphabetical order of spelling.

Afterthought

Science should involve doing something interesting ('HANDS-ON' as they say), with time for sharing opinions and thinking ~ thus, useful knowledge will arise that is also personally meaningful.

© Collins Educational Science Teachers — You may photocopy this page for use within the classroom.